Also by George Ortega

*Free Will: Its Refutation, Societal Cost
and Role in Climate Change Denial*

Exploring the Illusion of Free Will, Second Edition

CLIMATE RESCUE CAPITALISM:
USING PRODUCT PROFIT TO FIGHT CLIMATE CHANGE

GEORGE ORTEGA

A Happier World
White Plains, New York

Published in the United States of America
for A Happier World, White Plains, New York
by CreateSpace, September 2014

10 9 8 7 6 5 4 3 2 1
Climate Change
Ortega, George 1957

ISBN-13: 978-1501063114
ISBN-10: 1501063111

For humanity and our neighbors on planet Earth

CLIMATE RESCUE CAPITALISM:
USING PRODUCT PROFIT TO FIGHT CLIMATE CHANGE

TABLE OF CONTENTS

Introduction

The climate crisis that is upon us looms larger each day. As each year passes, and our world remains incapable of summoning the will to respond meaningfully, the crisis grows more menacing, and the cost of our eventual, inevitable response increases. This cost is measured not in millions, but in billions of dollars that could easily extend into the trillions. But that's not our primary concern.

As we pump more and more greenhouse gases into the atmosphere, our gravest threat is that the Earth's climate will change beyond a point after which global warming becomes unstoppable. If that happens, the human civilization we know will cease to exist within a few hundred years. A global human population expected to grow from our current seven billion to over nine billion by 2050 will thereafter be reduced to fewer than one billion, and that prediction is optimistic.

One would think our primary battle is with the climate, to be waged by physical scientists like climatologists, geologists, biologists and physicists. Yes, that battle must be won, but not before the battle we are currently engaged in is fought through and resolved. The arena on which this present battle is being waged lies not within the domain of the physical sciences. Its arena is the social sciences, with economics, sociology and psychology being the principle theaters of engagement.

The battle we wage is one between personal self-interest and collective self-interest, between the welfare of the few living in this generation and the welfare of the many, which in this case extends to generations not yet born. It is a battle between a proportionally miniscule percent of today's human population who control enough wealth to determine what does and does not get done about climate change, and the rest of humanity.

This controlling population — and I would estimate its numbers fall short of even one percent of our world's population — is incapable of doing what our climate crisis calls on them to do. They are incapable of this because they are human, and, as such, fall prey to the multitude of human failings and vices of which we are all too aware, and also subject. From a psychological perspective, humanity has no valid reason to expect that these controlling few will change their ways, at least before such moral reformation ceases to make any difference. And so, humanity's battle with them can be understood in very simple terms. Our best — and probably only — hope of winning is to deny them their one formidable weapon; money.

But because the political will does not exist to take that money from them through historical means like the 92 percent top U.S. Federal tax rate in place in 1959 under President Eisenhower, and the socialist experiments of Russia, China and other countries, humanity's hope must be pegged to a different, more universally acceptable, approach. That's what this book is about.

Just as the world's controlling population won that control by selling goods and services, humanity must use this same means to win back the level of political influence necessary to first winning its battle against them, and then winning the battle against climate change. Humanity must go into business for the sake and interests of humanity, and market products that compete with, and win against, products marketed by today's controlling population.

I refer to this market-based means of increasing humanity's ability to fight climate change as Climate Rescue Capitalism for two reasons. First, because it is a capitalist venture designed expressly to fund the fight against climate change. And second, because it is a free-market capitalist venture in every sense of the word. It requires no government participation, and infringes upon no one's personal, political or economic freedom. It is nothing more, or less, than the utilization of capitalist marketing principles and practices for the purpose of fighting climate change, rather than to further enrich the individuals and corporations whose stranglehold on our world's political will to fight climate change is neither ecologically sustainable, nor morally acceptable. Climate Rescue Capitalism is about waging economic warfare for the sake of the Planet. And although there will be some losers among our world's controlling population, even they stand to gain far more than they will lose by losing this battle.

A few words of advice to the reader: first don't be fooled by the simple and direct language that follows into imagining that the Climate Rescue Capitalism concept is in any way simplistic, or unsophisticated, or seriously limited in its potential. It is both as simple, and potentially powerful, a means of transforming our world's response to climate change, as Einstein's $E=mc^2$ equation was of transforming our world's physics and technology. Second, don't be put off by the reiteration of major themes and concepts. You are being called upon to see capitalism and the marketing of products in a profoundly different, and unique, way. Trust that this review is necessary to your understanding and appreciating Climate Rescue Capitalism's utility and implications.

If we accept its thesis and promise, what Climate Rescue Capitalism calls upon us to do should be as clear as it is obvious. Someone once had the thought that "if you build it, they will come." Climate Rescue Capitalism promises that if we market products on behalf of humanity's fight against climate change, shoppers will buy our products, and not only will we be helping save the human civilization we know, we will also be earning substantial compensation for our efforts.

The Climate Change Crisis

Our climate crisis received its first major thrust into public consciousness in 1988 when James Hansen, then head of NASA's Goddard Institute of Space Studies, appeared before Congress to warn them of the rapidly growing threat to human civilization being caused by dangerously high levels of greenhouse gases in the atmosphere. It was not until former Vice President Al Gore's Oscar winning documentary on global warming, *An Inconvenient Truth* premiered in 2006, however, that our world felt an informed sense of urgency regarding the threats posed by our changing climate. Unfortunately, that urgency was short-lived. Eight years later, Pew Research Center reported that Americans ranked global warming near the bottom of Presidential and Congressional priorities for the years 2009 through 2014, and that only 44 percent of Americans currently believe there is solid evidence the phenomenon exists, and is anthropogenic.[1] This is where we are now, and why our world desperately needs new, outside of the box, thinking and new approaches to dealing with our climate crisis. Before exploring how Climate Rescue Capitalism can help fund our fight against climate change, let us examine the crisis that faces us in more detail.

In 1988, the United Nations established the Intergovernmental Panel on Climate Change (IPCC) to assess the nature and extent of the climate crisis, and to recommend policies and procedures for its mitigation and adaptation. It is currently comprised of over 3,000 scientists from 194 countries, and has issued assessment reports on climate change in 1990, 1995, 2001, and 2007. Because its next published assessment is not scheduled for release until October 2014, we will consider IPPC findings from its 2007 Fourth Assessment Report (AR4), and previous reports.

The IPCC does not conduct climate research; its scientists evaluate evidence of climate change published in peer-reviewed scientific journals throughout the world. It is a consensus organization, which means that its published findings must be approved by the 194 member countries, with each country having veto power over the publishing of any and all specific findings. As they must gain acceptance by each of the largest greenhouse gas-emitting countries like the United States, China, and India, as well as each of the major oil producing countries, its assessments have been characterized as highly conservative.[2]

In 2007, the IPCC, then comprised of over 2,000 scientists from 154 countries, concluded that by 2100 the world's surface temperature is expected to rise by 3 degrees Celsius over its 1750 pre-industrial level.[3] It recommended that CO_2-eq emissions be reduced by 80-95 percent below 1990 levels before 2050 for a 50/50 chance of our averting the 450 parts per million (ppm) atmospheric CO_2 concentration level that would cause a surface temperature rise of 2 degrees Celsius over the pre-industrial level.[4]

According to a September 19, 2007 article in *The Independent*, (U.K.), the IPCC's conservative predictions for such a 2-degree Celsius rise include the following land mass consequences: [5]

- "Africa: Between 350 and 600 million people will suffer water shortages or increased competition for water. Yields from agriculture could fall by half by 2020.

- Asia: Up to a billion people will suffer water shortages as supplies dwindle with the melting of Himalayan glaciers.

- Australia/New Zealand: Water supplies will no longer be guaranteed in parts of southern and eastern Australia by 2030.

- Europe: Water availability will drop in the south by up to a quarter. Heat waves, forest fires and extreme weather events such as flash floods will be more frequent. New diseases will appear.

- Latin America: Up to 77 million people will face water shortages and tropical glaciers will disappear. Tropical forests will become savanna.

- North America: Economic damage from extreme weather events such as Hurricane Katrina will continue increasing.

- Polar Regions: The seasonal thaw of permafrost will increase by 15 per cent and the overall extent of the permafrost will shrink by about 20 per cent."

The effects of a 3-degree Celsius rise in atmospheric temperature, the expected outcome if our world meets only the pledges made at the 2009 Copenhagen conference on climate change,[6] include a 23-foot rise in sea level[7]. That 70 percent of the world's people reside in coastal plains[8] reveals the far-reaching impact such sea level rise would have on human civilization.

As dire as those environmental threats seem, they may pale in comparison to threats from geopolitical de-stabilization that could result as countries suffer collapsed economies, and battle over dwindling resources. In April, 2007, the U.S. Army War College's Military Advisory Board published a report titled "The National Security Implications of Global Climate Change," which concluded that "Many developing nations do not have the government and social infrastructures in place to cope with the type of stressors that could be brought about by global climate change...When a government can no longer deliver services to its people, ensure domestic order, and protect the nation's borders from invasion, conditions are ripe for turmoil, extremism and terrorism to fill the vacuum."[9] Without serious climate change mitigation and adaptation efforts to address these threats, the United States and other developed countries may fall prey to a climate change-driven eco-terrorism potentially more widespread and devastating than that waged by al-Qaeda, ISIS, and other radical terrorist organizations.

Many climate-driven events threatening political de-stabilization throughout the world are not merely forecasts; they are already happening. They include European desertification,[10] a world-wide surge in diseases, [11] rapidly shrinking coral reefs,[12] lower global staple-crop yields,[13] deadly heat waves,[14] epochal ocean acidification,[15] global increase in wildfires,[16] rapid glacial melting,[17] global ecosystem disruption,[18] accelerated permafrost thawing,[19] record tornadoes,[20] Amazon rainforest tree species composition changes,[21] ocean salinity changes,[22] weakening of the Atlantic Gulf Stream,[23] increasingly rapid sea level rise,[24] major phyto-plankton decrease,[25] and accelerating species extinction.[26]

Several months after *An Inconvenient Truth* premiered in 2006, Nicholas Stern, a former world bank economist commissioned by the British government to assess the economic impacts of climate change, published a 700 page report concluding that unless the world began soon to spend approximately one percent of our annual gross domestic product (GDP) on mitigation and adaptation measures, we would risk shrinking the global GDP by as much as 20 percent.[27]

That same year, British scientist James Lovelock, whose work in the 1960s helped established the foundation for the hypotheses that raising CO_2 levels could profoundly change the character of the Earth's climate, published his book *Revenge of Gaia* (after the Greek goddess Gaia personifying the Earth). Lovelock's assessment of the climate threat is encapsulated in the book's preface where he writes, "before this century is over, billions of us will die and the few breeding pairs of people that survive will be in the arctic region where the climate remains tolerable."[28] In his documentary, Gore placed Lovelock's warning on the deeply pessimistic end of scientific assessments, however ever-improving research methodology and findings are lending greater creditability to Lovelock's dire prediction.

In September 2006, The Associated Press reported that methane, which converts to CO_2 as it is released from thawing permafrost, is entering the atmosphere at a rate five times faster than previously expected.[29] In November 2006, Reuters reported that tropical peat bogs, which comprise about three percent of the Earth's surface and store 2 trillion tons of CO_2, had been completely omitted by scientists from the global greenhouse gas calculus.[30] In December 2006, *National Geographic News* reported that sea levels could rise 40 percent higher than previously predicted,[31] in part because Artic sea is melting twice as fast as expected by climate models.[32] In February 2007, *The Guardian* (U.K.) reported that the likelihood of ice sheet losses which would ultimately raise sea levels by 13 to 20 feet had gone from a 2001 assessment of "very low" to a 2007 assessment of 50/50.[33] And in May 2007, *The Christian Science Monitor* reported that the CO_2-absorbing capacity of continents and oceans had been greatly overestimated.[34]

The most ominous new development, that renders Lovelock's dire prediction unavoidable without an extraordinary amount of good luck and sustained international cooperation, also came in 2007. That year, James Hansen announced that the IPCC's Third Assessment Report seriously miscalculated as 450 ppm the level of atmospheric CO_2 concentration that our world must remain under by 2050 to avoid a Lovelockian future, revising the threshold to 350 ppm.[35] Bearing in mind that the

2007 CO_2 level was 385 ppm, that it is increasing at an annual rate of over 2 ppm, and that in April 2014, we surpassed the 400 ppm mark, Hansen's warning provides a powerfully clear picture of how huge a threat the climate crisis poses.

New developments since Hansen's 350 ppm warning reveal that the magnitude of the climate threat extends even beyond the ramifications of that revised benchmark. In April 2008, Reuters News Service reported that a third of the world's large ocean regions are warming two to four times as fast as previous average temperatures predicted.[36] And that same April, Nicolas Stern an-- nounced that his 2006 economic assessment under- estimated both the probabilities of temperature increases and the damage those increases would cost, and doubled from one to two percent of global GDP the amount it would cost to effectively ad- dress the climate crisis.[37] In June 2008, the Inter- national Whaling Commission reported that coastal oxygen starved zones have increased by one third since 2006.[38] Also that June, a study published in the British science journal *Nature* reported that the world's oceans had warmed 50 percent more over the last 40 years than previous research had de- tected.[39] Additionally, the Centre for Ice and Cli- mate at the Neils Bohr Institute of the University of Copenhagen highlighted the non-linear and poten- tially abrupt nature of climate change, and the threat we face by not doing enough about it soon enough, by reporting that Greenland ice core evidence reveals that approximately 15,000 years ago, global temperatures increased 5.5 degrees Celsius within 50 years.[40] In December 2009, NPR.org reported that the U.S. Pentagon has officially classified global warming a national security threat.[41]

Reviewing our political problem, the American People, who account for about 25 percent of annual global greenhouse gas emissions, have for the last five years ranked climate change at near the bottom of Presidential and Congressional priorities, and 66 percent of Americans currently deny there is solid evidence the crisis exists and is anthropogenic. Our world is seriously in need of a major game changer. Since it usually takes money to change public perceptions and governmental policy, the root of, and solution to, our climate crisis can be viewed in economic terms. Most specifically, how do we finance our fight against climate change? This book introduces a revolutionary way by which we can structure aspects of our capitalist economy to fund this fight.

Climate Rescue Capitalism

Climate Rescue Capitalism is a powerful vehicle by which business profit within a capitalist economy can be redirected in order to fight climate change. It is a way to direct product profits away from conventional corporations and toward the funding of the scientific, sociological and political initiatives needed to best mitigate and adapt to climate change. As such, it represents a pragmatic coming together of capitalist and socialist perspectives that maximizes the fundamental strengths of these two disparate economic systems, as they relate to climate change.

As a strategy, Climate Rescue Capitalism, is deceptively simple but of unparalleled promise. The idea is to create new companies we will call Climate Rescue Corporations that do business in order to finance the fight against climate change. These corporations would be owned and operated by private individuals as well as not-for-profit climate change organizations. They would manufacture products to compete with existing products offered by conventional companies. Consumers would choose Climate Rescue Corporation products over those of their competitors because they would prefer to see the profit from their purchases be used to fight climate change than to further enrich private individuals and corporations. For example, a Climate Rescue Corporation established to fund solar panel research markets margarine of equal value to margarine offered by conventional manufacturers. At the supermarket, a substantial percentage of consumers should select the Climate Rescue Corporation margarine over all others because they would rather see the profits from their purchase used to fund research on better and more cost-effective solar power technologies than for the personal gain of competing manufacturing company owners.

This selling point will work for virtually any now existing product that can be matched in quality, and sold for an equal price. There are literally hundreds of thousands of such products now in the marketplace. With every one of them, consumers would, virtually regardless of the specific climate change-related initiative to which Climate Rescue Corporations would donate their profit, buy these products over those of the competition. This should eventually translate into many billions of new dollars being channeled into the fight against climate change every year.

Climate Rescue Capitalism would change the nature of consumer spending in this and every other capitalist country in the world. Along with our taking the steps necessary to minimize the kind of wasteful materialistic culture that keeps greenhouse gas emissions at unsustainable levels, our current materialism would also become a primary funder of our decades-long climate change initiatives.

Purchasing Climate Rescue Corporation products will hopefully become so highly regarded a social institution that it would in many ways become taboo to not patronize them. Common sentiment would hold that anyone refusing to utilize this opportunity to help fight climate change, at no personal expense, through the purchase of Climate Rescue Corporation products, would be deemed indifferent and irresponsible. What good reason could anyone possibly offer for not buying Climate Rescue Corporation products that are equal in price and quality to competing products sold by conventional corporations? Given this rationale, Climate Rescue Corporations should succeed extraordinarily.

At this point, the reader may feel that Climate Rescue Capitalism seems too good to be true. How can Climate Rescue Corporations stay in business, and grow, if they give away all of their profit? We don't need to rely on a theoretical model or speculative predictions to answer that concern. In the next chapter, we'll begin to explore a working model for Climate Rescue Corporations that has existed and thrived for over thirty years.

Staying in Business While Donating all Profits: The Newman's Own Model

Newman's Own, a prototype profit-donating corporation established in 1982 by Actor Paul Newman and author Alex Hotchner, donates one hundred percent of its after-tax profit to charitable causes. To date, Newman's Own has donated over $400 million to thousands of charities, and has grown from offering one product to marketing a line of several dozen products. Currently, it donates $25-30 million annually to over 700 charitable organizations.

It would be mistaken to summarily conclude that Newman's Own is so successful because of Paul Newman's celebrity appeal. In fact, most celebrity owned and endorsed products fail in the marketplace. While no study has yet determined what factors contribute to the enormous success of Newman's Own, it is indisputable that Newman's Own has thrived and grown over the last three decades while donating one hundred percent of its profit to charity. However much of Newman's Own's success may be attributable to the celebrity of Paul Newman, it is nonetheless clear that selling products to fund important causes represents a viable, time-tested, and sustainable strategy.

We'll explore many details of the Newman's Own financial success and marketing plan later in the chapter devoted to developing a campaign to launch Climate Rescue Corporation products. For now, let's review the following key facts about Newman's Own's first 20 years of operation.[42]

In 1981, Paul Newman and his neighbor and business partner, Alex Hotchner each invested $20,000 to start their company. They launched their first product, salad dressing, in 1982, and within the first six months they earned $65,000 in profit from sales totaling $502,000. Their profit for the first year totaled $920,000

By 1984, they had sold 18,705,555 bottles of salad dressing and 8,371,726 jars of spaghetti sauce. A $25 million operation in 1985, their salad dressing earned a 3 percent share in a 600 million market, their spaghetti sauce owned 1.3 percent of a $770 million market, (Campbell Soup's Prego had a 30 percent share), and their popcorn held a 10 percent share in a $30 million market. By 1987, Newman's own was growing at a rate of about 20 percent per year, and in 1988 their gross earnings were $36 million. In 1989 Newman's Own earned a

16 percent pre-tax profit, which was about five times the 3.4 percent food industry average.

Expanding overseas to the United Kingdom, by 1993, their pasta sauce had a 2 percent market share, and their salad dressing had a 5 percent market share. In 1997, Newman hired his first professional COO, Tom Indoe, who discovered that fewer than 25 percent of Newman's Own consumers know that after-tax profits go to charity. By 2004, profits worldwide were increasing annually by over 10 percent, and Newman's Own was selling $190 million in products every year with only 18 employees. Virtually everything was outsourced, and they sold to 14 percent of U.S. households.

From the very beginning Newman's Own was a resounding success, clearly demonstrating that a business can thrive while giving all of its profit away. In the next chapter, we'll explore how their model can be used as a blueprint for implementing Climate Rescue Capitalism as a game-changing funding tool in our fight against climate change.

Implementing Climate Rescue Capitalism

Along with the powerful selling point of donating their profit to climate change mitigation and adaptation initiatives that include lobbying, and public education and mobilization, Climate Rescue Corporations would have another major advantage over conventional businesses. Free publicity from the media and from not-for-profit environmental organizations would translate to millions of dollars annually in free advertizing. Partly in response to public pressure, the Media would assume an important role in publicizing to consumers that Climate Rescue Corporations are being formed, and are offering products to fund the fight against climate change. Media corporations can be expected to not only introduce the concept of Climate Rescue Capitalism to the public, and urge consumers to buy Climate Rescue Corporation products, they would also very likely maintain an ongoing commitment to keeping the public informed about what new Climate Rescue Corporations are being created, what products they are selling, and which specific climate change-fighting initiatives would be benefitting from their product profits.

Media coverage should be capitalized on from the beginning. Once preliminary market research validates the success Climate Rescue Corporation products can expect to enjoy, work should begin on establishing as many Climate Rescue Corporations offering as many products as possible, in order to create a major media event. Were the process to start slowly, with one or two Climate Rescue Corporations offering one or two products, the media would not take much notice, and the advertizing potential of such media coverage would be minimal.

If, however, the Climate Rescue Capitalism concept is introduced to twenty or thirty entrepreneurs, philanthropists and not-for-profit environmental organizations, each agreeing to create a Climate Rescue Corporation, and together, agreeing to introducing their new products simultaneously, the media would pay much more attention to this orchestrated product-launch event and, of course, publicize it much more extensively.

Persuading individual Climate Rescue Corporation owners to work together in this way should not be difficult because the more publicity the concept and goals of Climate Rescue Capitalism receive the more products each individual Climate Rescue Corporation can expect to sell.

The best venue to begin selling Climate Rescue Corporation products may be supermarkets since so many of the products they sell are easy to duplicate, and feature packaging that can provide Climate Rescue Corporations an ideal way to inform consumers that all profits from the sale of their products are donated to a beneficiary climate change initiative.

Eventually Climate Rescue Corporation products could become so ubiquitous in supermarkets that they would be granted their own unique sections and shelf space. After consumers have been introduced to the concept of Climate Rescue Capitalism, and are routinely buying Climate Rescue Corporation products in supermarkets, Climate Rescue Corporations could very successfully branch out into other consumer product markets. Eventually there could even be retail stores exclusively devoted to selling Climate Rescue Corporation products.

Climate Rescue Capitalism could become such a prominent force in the retail product economy that in some markets, competition would no longer be between Climate Rescue Corporation products and conventional products, but rather between different Climate Rescue Corporations funding different climate change initiatives. Climate Rescue Capitalism can succeed in any capitalist economic system, and, over time, Climate Rescue Corporations could be established worldwide.

Climate Rescue Capitalism can also be a powerful mechanism for redirecting political influence on climate change policies and practices from rich individuals and corporations to the majority populations throughout the world. This would come about by the creation of Climate Rescue Corporations whose beneficiary organizations would lobby governments on behalf of majority populations and the cause of climate change mitigation and adaptation. In this way, our world's people would take a direct role in helping fight climate change by purchasing Climate Rescue Corporation products. No longer would the world's majority population be helping the rich become richer, and be further exacerbating climate change, with each product they purchase.

One reason Climate Rescue Corporations could eventually become a formidable force in many prominent product markets is that owning and operating Climate Rescue Corporations would be a profitable enterprise for the individuals and not-for-profit organizations who own them. While Climate Rescue Corporation owners would certainly not become as rich as owners of conventional companies, they could earn high enough salaries to justify their efforts in establishing and running a Climate Rescue Corporation. These salaries would be calculated as a percentage of the profit that is donated annually. For example, a Climate Rescue Corporation donating $5 million a year to a windpower climate change initiative would devote one million dollars per year to employee salaries and benefits. This donated-profit-to-salary proportion is based on the not-for-profit industry standard of devoting about 20 percent of revenues to operating expenses.

Since Climate Rescue Corporations can hire factories to manufacture, and brokers to distribute, their goods, very few employees are needed to run a start-up Climate Rescue Corporation donating millions of dollars a year to climate change initiatives. This assessment is based on Newman's Own having by 1987 donated over $15 million to charity, and being, at the time, run by only four employees. That means that four Climate Rescue Corporation co-owners could each earn $100 thousand per year running a Climate Rescue Corporation that donates as few as two million dollars a year to fighting climate change.

Individuals wishing to earn money by operating a Climate Rescue Corporation would not be the only potential owners. Some of us would establish Climate Rescue Corporations for purely altruistic reasons. For example, a philanthropist who originally considered donating one million dollars to a climate change not-for-profit organization could instead use that money to establish a Climate Rescue Corporation that would donate its profit to that organization. After a few years, his business could be donating far more than one million dollars each year to the organization, and the philanthropist will have helped that organization much more so than if he had made his original one-time contribution directly.

Celebrities would also be ideal owners of Climate Rescue Corporations. Having their name and picture on products would be great publicity for them, especially since all the profit from the products was being used to fund the very popular cause of fighting climate change. Celebrities would likely be among the first individuals to establish Climate Rescue Corporations since many of them have enough wealth to fully finance their Climate Rescue Corporation, and could start the business without incurring onerous financial risk.

The key to the success of Climate Rescue Capitalism lies in convincing individuals and not-for-profit climate change organizations to invest in establishing the initial Climate Rescue Corporations. This is where market research is essential. Implementing Climate Rescue Capitalism requires a small army of marketing professors and graduate students to research consumer acceptance of the Climate Rescue Capitalism concept. These academics could then use their research to introduce the concept and feasibility of Climate Rescue Capitalism to the academic world and to potential Climate Rescue Corporation owners. Once this is achieved, the next step would be for marketing professors and graduate students throughout the world to research specific aspects of Climate Rescue Capitalism such as which Climate Rescue

Corporation products would sell best, and which specific climate change initiatives would the public be most willing to support through their product purchases.

Gaining the participation of as many marketing professors and graduate students as possible would be a very effective way of providing potential Climate Rescue Corporation owners with free market research on their proposed products, thereby making it far less expensive for them to market their products. Marketing professors and graduate students would likely be eager to become involved with this research because Climate Rescue Capitalism is a unique marketing concept that has not been studied before, and because of the concept's world-changing potential.

Market research for Climate Rescue Corporation products would differ somewhat from market research for conventional products because the Climate Rescue Capitalism selling point of donating all profits to climate change-fighting initiatives would apply to all Climate Rescue Corporation products. Research into specific types of Climate Rescue Corporation products would also be applicable to similar products in the same market. For example, market research indicating that consumers would purchase a Climate Rescue Corporation's ketchup to fund greenhouse gas sequestration research would also implicitly show that the same consumers would also purchase relish, mustard, mayonnaise and other similar food products to fund biofuel technology research.

It is important to appreciate that donating all profits to climate change initiatives is what sells Climate Rescue Corporation products, and that once market research has established that this selling point is effective, all that needs to be determined is which existing consumer products fail to command unassailable brand loyalty, and focus Climate Rescue Corporation market research and production on these products.

Of course potential Climate Rescue Corporation owners will want to know which specific climate change mitigation and adaptation initiatives and organizations are most popular, and which products are most easily duplicated in quality, and sold for an equal price. This information would be determined by marketing professors and graduate students who would then make their research available to these potential owners.

In early September, 2014, I conducted a survey in White Plains, New York to determine if consumers would buy Climate Rescue Corporation supermarket products equal in price and quality to products they now buy to help fight climate change, and 44 of the 50 respondents surveyed answered "yes." The exact question I had respondents read so as to not verbally influence their answer was:

> If your supermarket offered new food products that were equal in price and quality to

the products that you now buy, and you knew that one hundred percent of the profit from these new products would be donated to the cause of fighting climate change, would you buy these new products?

Once market research documenting the success potential of Climate Rescue Corporation products has been introduced to philanthropists, entrepreneurs, celebrities, and not-for-profit environmental organizations, many other individuals will also likely decide to establish Climate Rescue Corporations from both philanthropic and personal profit motives.

After some years, when Climate Rescue Corporations are offering products worldwide in every segment of the consumer product market, what will have occurred is a revolutionary utilization of capitalism that allows for private gain and rewards personal initiative while also directly addressing the urgent need to fight climate change. Climate Rescue Capitalism represents a pragmatic coming together of our world's two historically competing economic ideologies, capitalism and socialism, in a way that maintains the efficiency and personal freedom of capitalism by maintaining the profit motive while concurrently directly funding our collective fight against climate change.

The success of Climate Rescue Capitalism lies in being able to convince many celebrities, philanthropists, entrepreneurs, other concerned individuals and not-for-profit organizations to establish Climate Rescue Corporations. These individuals and organizations must be shown that Climate Rescue Corporations can be realistically profitable for the owners, along with fighting climate change. Based on the standard not-for-profit operating-expense to direct-service proportion ratio of 20/80, devoting $200 thousand to salaries for every one million dollars donated to climate change-fighting initiatives should be accepted by consumers as fair, and not deemed exorbitant, compensation for owning and operating a Climate Rescue Corporation. If this turns out to not be the case, a more acceptable consensus could be reached as to what would be a fair and appropriate compensation for owning and operating a Climate Rescue Corporation that would also provide sufficient inducement for as many individuals and organizations to establish Climate Rescue Corporations as possible.

Notwithstanding, there remains, of course, the danger of company owners taking too great a percentage of net profits for salaries, thereby undermining the integrity of Climate Rescue Capitalism. It therefore seems prudent to establish a watchdog organization whose role is to oversee and regulate salaries and operating expenses of all Climate Rescue Corporations. This organization would assure consumers that exorbitant salaries were not being earned under the guise of donating profits to the fight against climate change. An official seal of approval that would appear on product labels could be granted to Climate Rescue Corporations conforming to this overseeing organization's delineation of a fair salary/operating expense structure, and consumers would be advised through the Media to purchase only those Climate Rescue Corporation products bearing their seal. An alternate way to assure fair salary/operating expense structures would be for Federal and State governments to set these compensation/operating expense ceilings.

It should be expected that Climate Rescue Capitalism will come under attack from conventional businesses that, unable to compete against the selling point of donating all profit to climate change initiatives, would band together in an attempt to get courts to rule that Climate Rescue Corporations enjoy an unfair advantage over conventional businesses. However, strong and broad public support, and various judicial and political strategies can ensure the right of these businesses to exist.

Climate Rescue Capitalism, utilized to fully harness the willingness among consumers throughout the world to fight climate change through the products they buy, could easily generate many billions of new dollars for many climate change research, political and educational initiatives each year. Lastly, an individual and/or organization will be needed to assume a leadership role in introducing Climate Rescue Capitalism to philanthropists, entrepreneurs, not-for-profit environ-

mental organizations, and to consumers throughout the world.

Climate Rescue Capitalism Theory: A Review

Before proceeding to the details of how Climate Rescue Capitalism can be launched globally, let us again explore the basic concepts involved. While this review may seem redundant, it is my personal experience that because the Climate Rescue Capitalism idea is so unique and revolutionary, it can easily be insufficiently, or incorrectly understood, and, therefore, invites being reiterated relative to various relevant contexts.

What is Climate Rescue Capitalism?

Climate Rescue Capitalism is an economic tool enabling product profit within any capitalist economy to be redirected toward the financing of climate change mitigation and adaptation initiatives. Climate Rescue Capitalism channels profits away from corporate owners and stockholders, and toward organizations and initiatives at the forefront of our fight against climate change. Climate Rescue Capitalism is very simple in theory, but it dramatically transforms how and why consumers buy products. It challenges philanthropic individuals and organizations as well as visionary entrepreneurs to establish and run Climate Rescue Corporations that sell products, and contractually bind themselves to donating one hundred percent of the net profit from these products to the fight against climate change. Climate Rescue Corporations essentially compete with conventional corporations for available consumer dollars by offering their own products.

What are the basic goals of Climate Rescue Capitalism?

Climate Rescue Capitalism is designed to empower our world to fight climate change in several ways:

- *Direct Funding of the fight against Climate Change:* Climate Rescue Corporations would directly generate substantial new revenue for fighting climate change.

- *Transferring Political Power from the Rich to the fight against climate change:* The fight against climate change is underfunded because rich individuals and corporations spend huge sums of money lobbying against this fight. As consumers re-channel their purchasing power away from products sold by rich corporations, and toward products sold by Climate Rescue Corporations, these rich corporations will have less revenue available to lobby against climate change mitigation and adaptation initiatives. At the same time, these initiatives will have substantial new revenue they can use to lobby for increased government support for the fight against climate change.

- *Empowering the World's People to Fight Climate Change:* Today when our world's people buy products, the rich individuals and corporations who sell these products become richer, while our climate becomes increasingly threatened. Climate Rescue Corporations can empower the world's people to fight climate change by offering shoppers throughout the world the opportunity to buy products whose profits are used to fund this fight.

- *Making Democracies more Democratic:* Climate Rescue Capitalism is also designed to make democracies much fairer by distributing political decision-making about climate change more equitably among all citizens. In most democracies, the few rich wield far more of this political power than the middle class and poor who make up the vast majority of voters in part because election laws allow rich individuals and corporations to directly and indirectly contribute huge sums to political campaigns. Climate Rescue Capitalism can empower Climate Rescue Corporations to sell products and donate their profits to political candidates who better support the fight against climate change.

How does Climate Rescue Capitalism work?

At retail stores, consumers would opt for Climate Rescue Corporation products over the same products offered by conventional for-profit corporations because they would prefer that one hundred percent of the after-operating-expenses profit from their purchases be used to fund climate change-fighting initiatives rather than make conventional corporations richer. For example, a popular not-for-profit environmental organization may create a Climate Rescue Corporation, and decide that it will donate all of its profits to initiatives conducting research on how best to communicate the urgency of the climate crisis to the general population. It markets a tomato sauce product that is equal in price and quality to the tomato sauces now offered by conventional manufacturers. Through product labels and other forms of advertising, the Climate Rescue Corporation informs consumers about their commitment to donate one hundred percent of the profit from each product purchase to that public outreach research. At supermarkets, consumers would choose this Climate Rescue Corporation's tomato sauce over other tomato sauces marketed by conventional corporations because they would rather see one hundred percent of the profit from their purchase be used to fund public education about climate change than be deposited into the bank accounts of competing company owners.

Will consumers buy Climate Rescue Corporation products?

Again, evidence for how successfully Climate Rescue Corporation products would compete with products offered by conventional corporations, comes from a survey conducted in White Plains, New York in early September, 2014. Asked the question, "If your supermarket offered new food products that were equal in price and quality to the products that you now buy, and you knew that one hundred percent of the profit from these new products would be donated to the cause of fighting

climate change, would you buy these new products?" 44 out of the 50 respondents surveyed
answered "yes." While some conventional corporation products command unassailable customer
loyalty, most can be very successfully competed
against by Climate Rescue Corporation products

How widely will Climate Rescue Capitalism's selling point work?

Climate Rescue Corporations' selling point of
donating one hundred percent of their profit to
climate change-fighting non-governmental organizations (NGOs) and initiatives should work for
virtually any existing conventional corporation
product that can be matched in quality and price.
There are at least hundreds of thousands of such
products in the marketplace. With each, consumers
would opt for the Climate Rescue Corporations'
products over those of the competition in order to
help fund the fight against climate change. This
would eventually translate into many billions of
new dollars being channeled to this fight every
year.

Will consumers value Climate Rescue Capitalism?

Consumers would value Climate Rescue Capitalism's strategy for using modern day materialism and spending to fight climate change. Climate Rescue Capitalism would enable hundreds of millions of consumers throughout the world to fight climate change each time they purchased a product from a Climate Rescue Corporation. Consumers would welcome the opportunity to buy products from Climate Rescue Corporations as a way of financially contributing to the climate change fight without it costing them any money to do so. After all, consumers would be buying the same quality products they would ordinarily buy, and paying the same price for these products. The only difference in their purchasing Climate Rescue Corporation products is that the profit from each purchase is used to fight climate change.

CLIMATE RESCUE CAPITALISM   45

Stages in Implementing Climate Rescue Capitalism

Establishing Climate Rescue Capitalism throughout the world as a more intelligent response to climate change than possible through conventional capitalism will be a monumental enterprise requiring extensive organization and collaboration. The essential stages for institutionalizing Climate Rescue Capitalism worldwide are

1. Concept Dissemination and Promotion
2. Organizational Leadership
3. Market Research
4. Climate Rescue Corporation Owner-Solicitation, *and*
5. Market Promotion

1. Concept Dissemination and Promotion

Establishing Climate Rescue Corporations throughout the world will require thousands of dedicated individuals who understand and appreciate the Climate Rescue Capitalism concept, and the plan for transforming Climate Rescue Capitalism from a high-minded concept to a working reality. This book is a first step toward disseminating and promoting Climate Rescue Capitalism. The next step is for climate change activists, NGO personnel, philanthropists, and journalists involved in the fight against climate change to refer potential participants to this book and its concepts, and to author and publish original articles and books describing Climate Rescue Capitalism and its goals to the international public. As this dissemination and promotion progresses, a leadership will naturally emerge to organize and implement subsequent stages of Climate Rescue Capitalism's development.

2. Organizational Leadership

While Climate Rescue Capitalism can certainly develop amorphously, a centralized organizational leadership can more effectively and efficiently establish Climate Rescue Capitalism worldwide. This leadership should be visionary, adept at thinking outside of the box, able to author highly sophisticated political and public communications, and be in possession of, or able to readily acquire, the personnel and financial resources necessary for leading collaborative efforts to institutionalize Climate Rescue Capitalism.

3. Market Research

In order to encourage organizations and individuals to confidently invest venture capital in new Climate Rescue Corporations, market research should be conducted to answer specific questions such as the following:

- How strongly and broadly does the public support the basic concept and goals of Climate Rescue Capitalism?

- Which markets (supermarkets, department stores, specialty stores, the Internet) will be best suited for introducing new Climate Rescue Corporation products? (I strongly recommend supermarkets.)

- Which specific products (salad dressings, canned goods, sauces, etc.) will consumers be most willing to buy from Climate Rescue Corporations?

- Which specific climate change-fighting initiatives will consumers be most willing to support?

How is This Market Research Best Acquired?

Climate Rescue Corporations hold a major advantage over competing conventional businesses with regard to market research. Since the concept of Climate Rescue Capitalism is of academic interest to the fields of economics, marketing, and advertising, extensive research into the above, and into other essential questions, can be conducted at no cost to prospective Climate Rescue Corporations by professors and graduate students at colleges and universities throughout the world. Also, research-for-donation contracts can be established between Climate Rescue Corporations that donate their profits to fund specific causes and specific colleges and universities.

For example a Climate Rescue Corporation supporting climate model research can contract with a university with a climate modeling center and a marketing department, and have the university's marketing professors and students conduct specific market research for the Climate Rescue Corporation in exchange for the Climate Rescue Corporations contractual commitment to donate its profits to the university's climate modeling center. Similar creative collaborations between Climate

Rescue Corporations and university economics, marketing, and advertising departments can generate, at no cost to new or prospective Climate Rescue Corporations, an extensive and detailed body of product-specific market findings that would otherwise cost tens of thousands of dollars if purchased from professional market research companies.

4. Climate Rescue Corporation Owner-Solicitation

Who will establish these new Climate Rescue Corporations?

Once a solid body of research exists strongly demonstrating that the public values the concept and goals of Climate Rescue Capitalism, and would purchase various Climate Rescue Corporation products supporting various climate change-fighting organizations and initiatives, philanthropists and not-for-profit organizations should be solicited to create new Climate Rescue Corporations. Philanthropists can be advised that rather than their making a large one-time donation to a favored climate change organization, they can establish a Climate Rescue Corporation that could ultimately generate far more, and continuing, revenue for that organization. Not-for-profit climate change-fighting NGOs should be encouraged to create Climate Rescue Corporation *arms* of their organizations that could generate more revenue than does their traditional strategy of soliciting donations. High profile politicians and sports and entertainment celebrities are also ideal candidates to establish Climate Rescue Corporations because their owning a successful Climate Rescue Corporation that contributes substantially to climate change-fighting initiatives will generate publicity promoting their professional careers and ambitions.

5. Market Promotion

How should new Climate Rescue Corporation products be introduced into consumer markets?

In certain industries and markets, conducting business independently, with minimal collaboration between Climate Rescue Corporations may be advantageous in various ways. To ensure the fastest and strongest success of individual Climate Rescue Corporations, however, extensive collaboration with other existing and prospective Climate Rescue Corporations is a smarter strategy. The media will pay little attention to new Climate Rescue Corporations sporadically entering various markets with their products. However, were ten, twenty, or more new Climate Rescue Corporations to orchestrate a simultaneous launching of new Climate Rescue Corporation products supporting various climate change fighting initiatives, media coverage would be far broader and longer lasting.

Adding to the newsworthiness of this simultaneous product introduction is the fact that Climate Rescue Capitalism is an unfamiliar socio-economic fund-raising vehicle bound to be controversial. Conventional businesses that fear losing sales revenues to these new Climate Rescue Corporations may attempt to brand the Climate Rescue Capitalism concept, and newly established Climate Rescue Corporations, as veiled socialism. As market research establishes that the Climate Rescue Capitalism concept and its goals enjoy strong and broad public support, such assaults can be confidently welcome, as they amplify the media attention individual Climate Rescue Corporations would receive.

To acquire extensive media coverage amounting to tens of millions of dollars in free advertising for new Climate Rescue Corporations, the Climate Rescue Capitalism organizational leadership should facilitate Climate Rescue Corporation collaboration in various ways such as the following:

- Climate Rescue Corporations should be encouraged to refrain from competing in markets against other Climate Rescue Corporations. Having prospective Climate Rescue Corporations collaborate in order to eliminate intra-Climate Rescue Corporation competition ensures the strongest success of each Climate Rescue Corporation.

- Climate Rescue Corporations should be encouraged to share marketing research findings; this is, of course, best accomplished when Climate Rescue Corporations are not competing with each other in various markets.

- Most importantly, an orchestrated introduction of as many new Climate Rescue Corporations selling as many new products, and donating their profits to as many different climate change-fighting initiatives and organizations as possible, will create broad media coverage which will be an invaluable asset to all Climate Rescue Corporations. Recipient climate change organizations should also be encouraged to concurrently send out mailings to their donor base informing them about the new products being sold whose profits will be donated to their organization.

The Climate Rescue Buy Aid Product Launch Campaign

Climate Rescue Buy Aid: *Using Product Labels to Mobilize Public Action on Climate change* is a product launch campaign that calls on the vast network of climate change-fighting NGOs to work together under the direction of a central organizing NGO in a campaign to market a line of 20-30 supermarket products worldwide. Marketing the products can generate substantial revenue for the climate change-fighting NGOs, and, the product labels can achieve Climate Rescue Buy Aid's second major objective; educating and mobilizing citizen action on climate change. Although an ambitious enterprise, Climate Rescue Buy Aid is actually far less difficult and risky than it might at first appear.

By recruiting celebrities to lend their names and images to each of the 20-30 products, and by activating the campaign through a widely promoted product launch event, marketing the products can be conducted as a virtually risk-free financial venture. The worst case scenario for the campaign would result in the sale of hundreds of millions of individual products, the label on each of them educating the public about climate change, while earning some revenue for the climate change-fighting NGOs. Profits from even a marginally successful campaign would easily cover all incurred costs, including the manufacturing, distributing and promoting of the 20-30 products.

Rationale and Benefits of the Campaign

It is widely acknowledged that public education and mobilization is the starting point and key to increasing the effectiveness of NGOs working to fight climate change. On their own, direct mail, TV and radio, print media, rock concerts, and Internet campaigns have thus far been woefully inadequate as strategies for sufficiently educating and mobilizing the public on climate change. An effective new strategy is desperately needed by climate change-fighting NGOs. The strategy this Climate Rescue Buy Aid proposal describes features the following benefits:

- It utilizes, as its information dissemination vehicle, a unique and bold approach; delivering its message to hundreds of millions of households throughout the industrialized world via the labels on the supermarket products that shoppers purchase routinely.

- Rather than costing climate change-fighting NGOs precious resources, as with mass mailings, or generating no direct revenues, as with donated media, the "product labels" strategy educates and mobilizes the public on climate change in a cost-free way that concurrently generates both direct and indirect revenue for the NGOs.

Rationale for the Produce Labels Approach

The idea of using product labels as a cost-free, revenue-generating climate change information dissemination vehicle evolved from four observations.

1. Supermarket products can be very easily and profitably marketed. Actor Paul Newman and author Alex Hotchner, two individuals with absolutely no food marketing experience, generated over $400 million in after-tax profits between 1982 and 2014 by marketing a line of *over forty successful food products* under the brand name Newman's Own.[43]

2. Newman's Own achieved its huge success while donating all $400 million in after-tax profits to thousands of charitable causes.

3. When asked in a survey if they would choose food products whose profit would be used to fight climate change over competing comparable brands, 44 out of 50 respondents answered "yes." Newman's Own did its own survey, without specifying a charitable cause, and reported a 76 percent "yes" response.[44] These two surveys indicate that the strategy of selling supermarket products to inform the public about, and generate revenue for, fighting climate change holds substantial promise.

4. As Newman's Own proved, the key challenge to successfully promoting food products can be easily met at virtually no cost through a product launch event attracting wide international press coverage.

Because Newman's Own has compellingly demonstrated the feasibility of marketing and promoting food products whose entire profit is donated to charitable causes, their business model and the details of their huge success are used throughout this proposal to describe the marketing and promotion features of Climate Rescue Buy Aid. Very fortunately, Paul Newman and his partner, Alex Hotchner, published a book in 2003 titled

Shameless Exploitation in Pursuit of the Common Good: The Madcap Business Adventure by the Truly Oddest Couple where they describe their business model.

Estimate of the Number of Products Climate Rescue Buy Aid can Sell During its First Year

During their first 29 months in business, Newman's Own sold 18,705,555 bottles of salad dressing.[45] Extrapolating from that figure an average annual sale of 7.7 million items, we can predict first year sales of over 7.7 million for *each* of the 20-30 Climate Rescue Buy Aid products marketed. Accordingly, if Climate Rescue Buy Aid decides to market 25 products, the campaign can expect to sell at least 193 million items during its first year.

The following additional factors should substantially add to the above base prediction for items sold during the first and subsequent years:

- The product launch event, promoted by not just one primary celebrity, as with Newman's Own, but with approximately 25 celebrities, will result in much wider press converge, and proportionately stronger sales.

- Newman's Own operated without a business plan and without paid advertising until 1997 when it hired Tom Indoe, a retired former executive with Del Monte and RJR Nabisco, as its first professional COO. During the subsequent six years, Newman's Own's line of salad dressings' market share increased from 2.1 percent to 5 percent.[46] By operating under a business plan, and devoting a portion of its budget to paid advertising from its onset, Climate Rescue Buy Aid can expect similarly larger market shares.

Financial Framework and Details of Marketing the Food Products

Ownership

This section describes options for who would actually market the Climate Rescue Buy Aid food products, and own the companies.

Option one is to follow the example of Newman's Own, and have each of the celebrities own their company. The celebrities would run their company like Paul Newman did, as an "S" corporation that must pay taxes, and donate all of its profit before December 31st of each year. Under this option, the celebrities would each provide the $98,000 (in 2014 dollars) in startup costs per product.

Under option two, a group of about 20-30 climate change-fighting NGOs would *each* create and run a food company, and use their tax-exempt status to earn higher revenues than possible through an S corporation. With this option, the celebrities would lend their name and image to the packaging, take part in the product selection process, and attend the product launch event. Along with the tax benefit, this option would provide the advantage of granting each NGO full control over how its earnings are used. This control would, naturally, enable each NGO to apply those earnings to its own programs.

Option three would be best for climate change-fighting NGOs with limited finances. Under this option, a group of several of these NGOs would partner together so that the start-up financing would be affordable to each of them. They would, together, agree on how to divide costs, how profits would be distributed among the group, and how product label space would be apportioned.

Under Option four, a single climate change-fighting NGO would secure the entire financing for the 20-30 food products, and determine exactly how the product profits would be used.

With each of these options, the campaign would be planned, coordinated and implemented by one central NGO. This organized collaboration would streamline, and save money on, such details as product selection, packaging and label design, and campaign promotion.

Start-Up Financing

In 1982, by forgoing expensive market research, conducting in-house product testing, outsourcing manufacturing and distribution, and using no paid advertising, Newman's Own launched its salad dressing with an investment of $40,000.[47] Using the Newman's Own model, in 2014 dollars, it costs about $98,000 to market a similar food product.[48] For 25 Climate Rescue Buy Aid products, that would represent an investment of $2,450,000. Depending on the Climate Rescue Buy Aid leadership's ability to attract volunteers to help promote the campaign, additional costs should amount to no more than an additional one million dollars.

Product Selection

Although Paul Newman personally chose which products to market, it is recommended that an industry professional be consulted to determine which product categories provide the greatest opportunity for new brands, and which products are most easily replicated in price and quality.

Product Testing and Market Research

Paul Newman, a home gourmet, avoided expensive product testing and market research costs by personally approving the taste and ingredients for each of his products, and by conducting his own in-house product vs. competition taste-testing.[49] Given his success with over forty products, and the results of two surveys suggesting an 88 percent and a 76 percent public willingness to opt for climate change-fighting products over those of competitors, it is recommended that Climate Rescue Buy Aid conduct this phase of the campaign following the Newman's Own example of minimal market research expenditures.

Product Manufacturing and Distribution

Outsourcing the manufacturing and distribution of the products, as is done by Newman's Own, is also recommended for Climate Rescue Buy Aid products in order to minimize start-up costs. The only recommended shift from the Newman's Own model would be to, after a period of two to three years, consider forgoing the loyalty Mr. Newman maintained toward his manufacturers, and establish manufacturing plants for Climate Rescue Buy Aid products as a way of maximizing profits.

*Significance and Organization of the Product
Launch Event*

Products judged by shoppers as inferior in
quality to competing products routinely outsell
their competitors because of effective advertising
and promotion. Following Newman's Own example, although Climate Rescue Buy Aid food products should be processed to insure the highest
quality, an effective promotion campaign like the
product launch event conducted for Newman's
Own's salad dressing will ensure that Climate Rescue Buy Aid products compete aggressively with
the products of food industry giants.

For their product launch, Newman and Hotchner rented out a small bar in upper Manhattan for
one evening, and staged a minor media event.[50]
They invited reporters and camera crews from all
of the New York newspapers, and from the Associated Press. The event immediately received
inter-national coverage, and within days Newman's Own was inundated with orders from supermarkets like Shopwell, A&P, Stop & Shop and
others.[51] By staging a publicity event like the one
Newman and Hotchner staged, Climate Rescue
Buy Aid can bring their line of 20-30 food products
into supermarkets throughout the world.

Celebrity power is the key feature of this campaign. Paul Newman attracted enough publicity
from his small event to sell 7.7 million bottles of
salad dressing during the subsequent twelve

months. One can easily imagine how much more publicity and sales 20-30 celebrities would generate for Climate Rescue Buy Aid products from a larger event launching 20-30 food products.

Product Marketing Details and Sales-Profit Projections

Because each specific market is unique, different Climate Rescue Buy Aid food products would succeed to varying degrees against different competitors. Using Newman's Own's first product, salad dressing, as an example reveals how well Climate Rescue Buy Aid products can be expected to succeed.

Example: Newman's Own Salad Dressing

In 1982, with a total investment of $40,000, Newman's Own introduced its first product, salad dressing. Commanding less than one percent of a market worth $600 million, it earned $3,204,335 in gross sales during its first year.[52] By outsourcing manufacturing and distribution, and forgoing paid advertising, Newman's Own's dressing earned an after-tax profit of $397,000 during that year.[53] This profit, amounting to 12.5 percent of gross sales, is about five times the food industry profit standard of 3.4 percent of pre-tax sales.[54]

Projection: Climate Rescue Buy Aid's Salad Dressing

Using the Newman's Own model, it would cost $98,000 in 2014 dollars to market a salad dressing. If this dressing commanded 0.5 percent of the current market worth $2 billion,[55] it would earn $10 million in gross sales. Implementing the Newman's Own strategy for manufacturing, distribution and promotion, Climate Rescue Buy Aid's product would earn an after-tax profit of $717,000 at 12.5 percent of gross sales, or an after-tax profit of $180,000 at the industry profit standard of 3.4 percent of gross sales, during its first year.

Based on Newman's Own's salad dressing success, we can establish a proportional sales/profits projection for a line of 25 Climate Rescue Buy Aid products. If each of the 25 Climate Rescue Buy Aid food products earned a 0.5 percent share of product markets with annual sales of $2 billion, their combined gross sales would amount to $250 million. After-tax earnings for the 25 products would amount to $17.9 million using a 12.5 percent-of-gross-sales profit margin, and $4.5 million using a 3.4 percent-of-gross-sales profit margin. These figures highlight that even with a marginal market share of 0.5 percent, (as contrasted with the 88 percent and 76 percent market share potentials suggested by the two shopper surveys) the 25 Climate Rescue Buy Aid food products would do phenomenally well.

Conclusion

Human civilization is in danger of collapsing. The stakes could not be higher. What are we willing to risk? What changes are we willing to undertake in order to provide humanity with a future worthy of the name? Seen relative to the major changes in our lifestyle that the urgency of the climate crisis may compel, like forgoing travel, and doing without air conditioning and heating, and paying much higher taxes, Climate Rescue Capitalism does seem like much of a sacrifice.

What would this revolutionary model for funding the fight against climate change cost us? Would we have to settle for poorer quality products? No. Would we have to pay higher prices for these products? No. In truth buying more of our products

from Climate Rescue Corporations entails no sacrifice from us at all. All we would be doing is buying
the products we now buy from different sellers.

To the contrary, these Climate Rescue Corporations would provide us with the opportunity to do
something very good for ourselves and for our
world's future generations every time we buy one of
their products. And we would feel very good about
doing this good. We would also feel good about
learning more and more about climate change by
reading with interest and motivation the routinely
updated information provided on the package labels
of the products we buy. These small joys are, of
course, insignificant when compared to the good we
would be doing to hasten our world's meaningful
response to the climate crisis. This is Climate Rescue Capitalism's real gift.

We can now only vaguely imagine what form
our progress in fighting climate change will take as
hundreds, and then thousands, and then tens of
thousands of new Climate Rescue Corporation products emerge in our global marketplace. How will
our network of climate change-fighting organizations spend the millions, and then tens and hundreds
of millions, and ultimately billions of dollars these
Climate Rescue Corporations will generate for
them? How quickly will the profits from our product purchases translate into a strong and resolute
political will to do what we can do — what we must
do — on behalf of our planet?

One might become fearful of the changes in our
world that Climate Rescue Capitalism could set in
motion. Clearly, the fight against climate change is

not our world's only underfunded cause. How soon will it be before medical researchers and global poverty organizations begin to create their own profit-donating corporations to fund these also very important causes? And how long will it be before shoppers are faced with the moral dilemma of having to choose between one product whose profit fights climate change, and a competing product whose profit fights cancer, or illiteracy, or injustice?

We can be sure that as Climate Rescue Corporations extend their reach into more and more product markets, elements among us will rise up to proclaim that these new corporations and the products they sell are the work of the devil, or a reincarnation of communism. Whether they act from self-interest or simple ignorance, Climate Rescue Capitalism and the corporations and products it gives rise to will have its share of detractors.

But do we truly have any choice other than to go forward with this potentially world-saving socioeconomic experiment? Can we in good faith decide that Climate Rescue Capitalism is so unorthodox and radical a utilization of our free-market capitalist system that we will instead opt for waiting until Mother Nature, with uncharacteristically brutal resolve, forces our hand on climate change? Can we in good faith continue to idly wait for her too-horrible-to-imagine assault against us. It may come through a devastating category three or four hurricane that directly hits, and cripples, New York City, potentially bringing the entire U.S. economy to a grinding halt. One can imagine even greater punishments like global pandemics that result from ever-rising surface temperatures on the Planet.

No, we cannot afford the consequences of leaving humanity's future in nature's hands. We must opt for first engaging in the more winnable battle against those in control of what our world does and does not do about climate change. If we don't wage this economic battle — if we don't accept and use the gift of Climate Rescue Capitalism as a way to win this battle against those few among us who are preventing our world from waging battle against our changing climate — then surely we have lost, and are lost.

The time to prepare ourselves to wage this battle for human civilization and for our neighbors on the Planet is at hand. The opportunity Climate Rescue Capitalism affords us to begin this fight in earnest stands before us. Reason offers no viable alternative. The time has come to establish Climate Rescue Capitalism as a means — and perhaps our only viable means — of providing humanity with the chance of a sustainable future. May goodness, and reason, and luck guide us through this grand and bold new enterprise.

Endnotes

[1] Pew Research Center (2014). Climate Change: Key Data Points from Pew Research: The American public routinely ranks dealing with global warming low on its list of priorities for the president and Congress. This year, it ranked second to last among 20 issues tested.

[2] Melting ice means global warming report all wet, say some experts. *USA Today*, January 28, 2007. (Note: many of the following citations were culled from The Heat is Online's archives.)

[3] Harvey, Fiona. Climate scientists warn of chaos from human action. *Financial Times*, February 2, 2007, as cited by EcoEarth.Info.

[4] Zhou, Maggie. What's the right target - 350 or 450 ppm? *Secure Green Future*, September 17, 2009.

[5] Milmo, Cahal. 'Too late to avoid global warming,' say scientists. *The Independent*, September 19, 2007.

[6] Climate pledges would still mean 3 degree rise: U.N. *Reuters*, December 17, 2009.

[7] Grice, Andrew. 3 degrees: Chief scientist warns bigger rise in world's temperature will put 400 million at risk. *The Independent*, April 15, 2006.

[8] Greenpeace. Sea Level Rise. July 4, 2012.

[9] Eilperin, Juliet. Military sharpens focus on climate change; a decline in resources is projected to cause increasing instability overseas. *The Washington Post*, April 15, 2007.

[10] Brown, Paul. Sahara jumps Mediterranean into Europe; Global warming threatens to create dust belt around the globe. *The Guardian*, December 20, 2000.

[11] Epstein, Paul. Is global warming harmful to health? *Scientific American,* August 2000, accessed from Mindfully.org.

[12] Evans, Dominic. Coral reefs are shrinking fast - UN report. *Reuters*, September 12, 2001, accessed from OceanConserve.org.

[13] Brown, Lester, R. World facing fourth consecutive grain harvest shortfall: wheat and rice prices moving up. *Plan B Updates*, Earth Policy Institute, September 17, 2003.

[14] Bhattacharya, Shaoni. European heatwave caused 35,000 deaths. *New Scientist*, October 10, 2003.

[15] Hecht, Jeff. Alarm over acidifying oceans. *New Scientist,* September 25, 2003.

[16] Huebner, Al. Burning Earth: Linking wildfires to global warming. Toward Freedom. October 8, 2007,

[17] Kirby, Alex. Kazakhstan's glaciers 'melting fast.' BBC News, September 4, 2003.

[18] Cropley, Ed. Global warming hits species all over the world. Reuters, March 28, 2002, accessed at Biodiversity & Human Health.

[19] Calamai, Peter. Tundra test stuns scientists; Carbon dioxide could be dumped into atmosphere. Raises spectre of accelerated global warming. *The Toronto Star*, September 23, 2004.

[20] NOAA reports record number of tornadoes in 2004. *Science Daily*, January 5, 2005.

[21] C02 wreaking havoc with Amazonian forest. *Agence French Presse*, March 11, 2004, accessed at *theage.com*.

[22] Atlantic's salt balance poses threat, study says. *The Toronto Globe and Mail*, **December** 18, 2003, accessed at theheatisonline.org.

[23] Sample, Ian. Alarm over dramatic weakening of Gulf Stream. *The Guardian*, December 1, 2005.

[24] Boyd, Robert S. Sea levels rise fast: 1 inch in 10 years. *The Seattle Times*, July 10, 2005.

[25] Borenstein, Seth. Warmed-up oceans reduce key food link. Associated Press, December 6, 2006, accessed at *The Washington Post*.

[26] Analysis: Global warming killing some species; Up to 200 species, including penguins and polar bears, are in big trouble. Associated Press, November 21, 2006, accessed at msn.com.

[27] McSmith, Andy and Brown, Colin. Climate change: US economist's grim warning to Blair's Cabinet. *The Independent*, Friday, October 27.

[28] James Lovelock, *The Revenge of Gaia*. Basic Books, New York, 2006. p. xiv

[29] Methane escaping five times faster than previously thought, study says methane a new climate threat. The Associated Press, September 7, 2006, accessed at theheatisonline.org.

[30] Tropical peat bogs stoke global warming-report. Reuters, November 6, 2006, accessed at planetark.org.

[31] Roach, John. Sea level may rise 40 percent higher than predicted, study says. *National Geographic News*, December 14, 2006.

[32] Black, Richard. Arctic melt faster than forecast. BBCNews.com, April 30, 2007.

[33] Adam, David. Climate change: scientists warn it may be too late to save the ice caps. *The Guardian*, February 19, 2007.

[34] Spotts, Peter, N. Nature's carbon 'sink' smaller than expected; Earth in 2100 could be up to 2.7 degrees F. hotter than previously predicted, studies say. *Christian Science Monitor*, May 3, 2007.

[35] Beck, Amanda. Carbon cuts a must to halt warming-US scientists. Reuters, December 13, 2007.

[36] McCool, Grant. Warming trends rise in large ocean areas: study. Reuters, April 9, 2008.

[37] Adam, David. I underestimated the threat, says Stern. *The Guardian*, April 18, 2008.

[38] Beeby, Rosslyn. Oxygen-starved oceans rapidly dying. *The Canberra Times*, June 25, 2008.

[39] Hood, Marlowe. Oceans warming faster than realized. *Discovery News*, June 18, 2008.

[40] Greenland ice core analysis shows drastic climate change near end of last ice age. *Science Daily*, June 19, 2008.

[41] Gjelten, Tom. Pentagon, CIA Eye New Threat: Climate Change. National Public Radio, December 14, 2009.

[42] Paul L. Newman and Alex E. Hotchner, *Shameless Exploitation in Pursuit of the Common Good: The Madcap Business Adventure by the Truly Oddest Couple*. Doubleday Publishing, New York, 2003, pp. 202-5.

[43] Newman's Own

[44] The Soul Brand. *Reveries*. March 1999.

[45] Newman and Hotchner, *Shameless Exploitation*, p. 204.

[46] Jon Gertner. Newman's Own: Two Friends and a Canoe Paddle. *New York Times*, November 16, 2003, sec. 3., p. 4.

[47] Newman and Hotchner, *Shameless Exploitation*, pp. 47-49.

[48] Based on The Consumer Price Index.

[49] Newman and Hotchner, *Shameless Exploitation*, p. 33.

[50] Ibid., p. 53.

[51] Ibid., p. 59.

[52] Newman and Hotchner, *Shameless Exploitation*, p. 97.

[53] Ibid. (Note: This figure may be low; on p. 204, first year profit is reported as $920,000).

[54] "Newman's Own, Inc. *International Directory of Company Histories*, Vol. 37. St. James Press, 2001. Reproduced in Business and Company Resource Center. Farmington Hills; Mich.,: Gale Group. 2006.

[55] Bruce Horowitz. Chocolate Chip Salad Dressing. *USA Today*, May 12, 2014.